METRO

METRO

George Szirtes

Oxford   New York
OXFORD UNIVERSITY PRESS
1988

Oxford University Press, Walton Street, Oxford OX2 6DP

Oxford New York Toronto
Delhi Bombay Calcutta Madras Karachi
Petaling Jaya Singapore Hong Kong Tokyo
Nairobi Dar es Salaam Cape Town
Melbourne Auckland
and associated companies in
Berlin Ibadan

Oxford is a trade mark of Oxford University Press

First published 1988

British Library Cataloguing in Publication Data
Szirtes, George
Metro
I. Title
821.' 914   PR6069.Z7
ISBN 0–19–282096–6

Library of Congress Cataloging in Publication Data
Szirtes, George, 1948–
Metro.
I. Title.
PR6069.Z7M48 1988   821'.914   87–31271
ISBN 0–19–282096–6 (pbk.)

Set by Wyvern Typesetting Ltd.
Printed in Great Britain by
J. W. Arrowsmith Ltd., Bristol

For my Father at Seventy
(a photograph aet. 35, early one summer)

Who would have thought that the slim dark man
with the short moustache, playing cards
with friends in a country garden would be
living at seventy on a foreign pension?
No one then. The summer is dutifully
photogenic. He wears the trace of a tan
and only his eyes betray the faintest hint of tension
as if they could look both backwards and forwards.

ACKNOWLEDGEMENTS

SOME of these poems first appeared in *Encounter, The Honest Ulsterman, The Jewish Chronicle, The London Magazine, Margin, Poetry Review, Poetry Voice, The Spectator,* and *The Times Literary Supplement.*

CONTENTS

THE LUKÁCS BATHS

1

It's circa 1900 and five women
have gathered here in semi-darkness
prepared to prophesy their own extinction.
The water shimmies down a pebbled wall,
a fountain hesitates. Their swimming costumes
are wasps' nests soaked through, softened by the gush,
their bathing caps are a green efflorescence.
They are the light at the bottom of deep pools
wobbling in uncomfortable sunshine
with rheumatic feet, imagining a Greece
ravaged by wars, prepared, they say, to sink.

2

Inside every grandmother there sits
an attractive young girl mouthing pieties,
complaining of sore lips or God knows what.
They prophesy the past with unerring accuracy;
history for them is painful gossip
half way between myth and memory.
They are on nodding terms with skeletons
who take the shape of husbands in dull rooms,
and they can tell the future as it shrinks
into its faint determined pattern.
It's hard to like them, harder to dislike them.
Their faces are light wrinkles in the water.

3

An enormous beech is jutting from the yard.
The walls, just as in crematoriums
are stuck with plaques in a handful of languages.
My shoulder's better. I can move my leg.
God bless these healing waters. I can walk.
Inside and on the roof the swimsuits bulge.
I'm watching two old women as they swim
and push away the past like tired waves.

THE HOUSE DREAM

I dreamt of a house in which a man was killed.
He loved the yellow bones of houses and tall windows.
Their shining bodies stood about him. He was tiny,
a child at under five foot two. His hands were minnows.

An empty niche dimpled the wall outside,
a portrait of his buxom wife hung vastly
in the biggest room. His desk had twisted legs.
There was about him something smelly and ghastly—

in retrospect at least, knowing him dead
and horribly so. His old housekeeper wept
among the statues. Where he wrote his slight poems
the wife of a drunken architect now slept.

And yet the house was beautifully at rest,
the mansard like a tight cap on his head,
a vague greenness moved and grew outside
and flowers sung gently in the flowerbed.

My dream or his? I thought I'd like to leave
some tiny vestige of self, collateral,
a History of Since Then, perhaps a poem
to celebrate the peaceful and unnatural.

I knew the name beyond my face and saw it
enter the hall and wipe its foot on the scraper.
The wind was in the gutter and the light
was flaking off the walls like painted paper.

A CARD SKULL IN ATLANTIS

The *Atlantis Paper Co.* to be precise—
purveyors of artists' materials from their warehouse
in Garnet Street, a stone's throw from the river,
vendors of paint, small bottles, aerosols,
but paper chiefly—cream, pale oatmeal, speckled,
translucent, edges stiffly cut or deckled.

A pirouette of spectacular bones blew softly
in the draught of the door. You could construct
the skeleton by cutting and glueing together
the pages of a book with delicate labels:
the thing required patience to collate,
life-sized at last, if rather underweight.

There is a crystal skull, I do believe,
in the British Museum, more articulate,
more valuable perhaps, but card will do;
it grins among the sketchbooks with its patent,
its ethmoid, larmical and zygomatic,
it could even whisper something politic

of skulls like paper, piled high in ditches,
two sets of grandparents, an uncle or two,
of cousins boarding trains, securely labelled,
and people watching one another from windows.
Under the eyes their bones flare for a minute,
collapse to powder on a distant planet,

subside and sink, and form a kind of silt,
washed down by rivers, drifting among boots
and waste from sewers, sticking to dead branches,
caught by the ebb tide sun along the bank,
sucked finally to sea in salt-sour smells,
and settling dumbly among rocks and bells.

GRANDFATHER IN GREEN

My grandfather, the Budapest shoemaker
    wrote plays in his spare time, and then he died.
His body became a pebble on a beach
    of softness across which swept the pale green tide.

Pale green, I think, would suit him as a tint—
    under his eye, or thinly flexed across
the hooked bridge of his nose. His sour complexion
    was cooking apples, a summary of loss,

each a pucker in the flesh. His waistcoat
    was grey as clouds, a pale green handkerchief
blossoming from the pocket. Even his tongue
    would sit in his mouth, soft and green as a leaf.

And so he returned to nature after all,
    the pale green gall within him in the shut
cavern of his stomach, and the green
    smell of gas still lingering in the hut.

SIREN VOICES

The sirens sang it: it was the old song
of Europe and the philanderer, his moustaches
brushed and waxed, a sly wink of galoshes
on the pavement as he strolled along.

The sirens sang and held the world together.
They sang of Babel in a tangle of veins,
of Viennese neurotics, Puritans
in Haarlem, and they sang of Goethe's mother.

They praised the classical style which makes good ruins,
they praised French gardens and Lord Burlington's villa,
they praised the stocks, the guillotine, the colour
of blood in ditches, washed away by rain.

They praised the colonists who reaped and sowed
rich whirlwinds, and they also sang out loud
the virtues of the sad and graceless crowd
whose every bone and button would corrode.

There was a verse on Simplicissimus,
another on Candide. They sang the wrecks
of ships and states, of innocence and sex,
the continent's incurable messiness.

They sang the jaundiced houses and the sky
trapped in a window, the clouds locked in a pool,
sang Dostoevsky and the Holy Fool
who's doomed to suffer and, at last, to die.

The sirens sang all this and strummed their harps
with barely a tear. Their voices might have been
the sea sprayed up, a light necklace of green,
a brilliant chromatic play of sharps.

ON A WINDING STAIRCASE

I climb these stairs which might be by Vermeer.
Light drools like spittle from the rails. They wind
towards a window, and even from down here
I make out the faint iron bands that bind
the house together in one act of will.

The writing on the wall says *Carpe Dym*
attached to the name of a national hero, or
it could well be the local football team.
Upstairs are voices I have heard before
that hook and draw you up as on a line,

to something cramped, imprisoned and defined
by yards and corridors. I run my hand
along the wall and feel it sweat and grind
its teeth. A brilliant light is in command,
a fist of light within an iron frame.

One purpose, one cohesion. People spill
from monumental gateways, accidents
of sun and shadow, leaving at their peril
the fortress of controllable events,
venturing out and over the world's rim.

I wait to see a family descend
down thirty years, each of them framed alone
before the window as they comprehend
the force that welds them with the light in one
unbreakable and static composition.

They owe a debt to history, that calm
and droning music which slows to a dead march.
Look at these maps, as wrinkled as a palm,
repoussoir instruments, an antique arch,
a girl with a trumpet, bent on playing Fame.

METRO

'What should they do there but desire'
DEREK MAHON

METRO

1 At my aunt's

My aunt was sitting in the dark, alone
Half sleeping, when I crept into her lap.
The smell of old women now creeps over me,
An insect friction against bone
And spittle, and an ironed dress
Smoother than shells gathered by the sea,
A tongue between her teeth like a scrap
Of cloth, and an eye of misted glass,
Her spectacles with the image of a lit room
Beyond the double doors, beyond the swing
Between the doors, and my head in her bosom
At rest on soft flesh and hard corsetry,
And in that darkness a tired and perfumed smiling.

*

Across the city darkened rooms are breeding
Ghosts of elderly women, nodding off
Over the books their grandchildren are reading,
Or magazines or bibles or buttons to be sewn,
With letters, patterns, recipes, advice.
Some of them might have the radio on
Like her, my aunt, who will remain alone
Within that room in which I visit her,
Ascending to her skin, which is rough
About the mouth, with hard nodules, like rice,
(Her face glows like a lantern) and she says
There is a God, the God of the Jews, of Moses and Elias,
But this is not the time to speak of him.

*

And here my aunt is happy, and her sister,
Both happy in their roles. And the child
Is happy in the reading of a tale
That ends in triumph over the wild
Succubi of his imagination:
The dwarfish furies of the forest, the lank
Raincoated ghosts who pester
The living daylights out of night,
The stepmothers who live beyond the pale.
The city waits like an armchair. A slight
Woman sits there, watching, as the evenings shrink
About her, and the city opens its arms
And welcomes her to its administration.

*

There are certain places healthy to have lived in:
Certain streets, hard cores of pleasure:
Their doorways are ripe fruit, stay soft and open,
Exhaling a fragrance of drains or tobacco,
Others are more proper, starched and sun-eaten,
Doorways where things happen
In a particularly fortunate way, which echo
To words of parting, or thrill to an exact measure
Recollected in the pleat of an arch;
Doorways which see military bands march
Across a square on a blazing hot afternoon,
Or catch a particular angle of the moon.
There are places to be happy in if only you can find them.

*

The Metro provides a cheap unending ride
If you switch trains below the city.
There is a whole war to be fought out under
The pavements; I can hear the faint thunder
Of artillery in Vörösmarty Square:
The cobbles shake, move gently from side to side
With microscopic accuracy, and ice creams
Wobble in their goblets. The cavity
Beneath the streets is filled with the blare
Of surface traffic. The city is all dreams
And talk, and rumours of talk. The place below,
Is treacherous. You don't know
Who your friends are, who you are yourself.

*

It is everything that is past, the hidden half,
A subcutaneous universe in which
Our fate is to be the dramatis personae
Of geographers who place us more precisely
Than we can ourselves. I place a woman
On a train and pack her off to Ravensbruck:
I send out a troop of soldiers to summon
The Jews of this fair city.
                              Off she goes,
Repeating her unknown journey, and I must look
To gauge the distances between us nicely.
I see a voice, the greyest of grey shadows.
Lead me, psychopompos, through my found
City, down into the Underground.

*

2   *Undersongs*

I love the city, the way it eats you up
And melts you into walls along with stone
And stucco till your voice assumes a tone
As crinkled, crenellated, creviced as itself,
And you can recognize it in a shop
Like something heard through windows. Human forms,
Detail and allegory: the twelve
Months of the year, the forces of nature, or
A frieze of leaves and angels where the worms
Have eaten away the substance of a voussoir
Above a flaking open door,
A spray of lace or foam, a mouchoir
In plaster flung at the late empire.

*

The empire underground: the tunnelling
Begins, the earth gives up her worms and shards,
Old coins, components, ordnance, bone and glass,
Nails, muscle, hair, flesh, shrivelled bits of string,
Shoe leather, buttons, jewels, instruments.
And out of these come voices, words,
Stenches and scents,
And finally desire, pulled like a tooth.
It's that or constancy that leads us down
To find a history which feels like truth.
The windows cannot speak because we pass
Before them all too often but the bricks know
What they stand on. There is no town below,

*

It's only bits and pieces, as above.
You have to watch your language though: the words
Are muddy, full of unintended puns
And nervy humour. Waking afterwards
You feel soiled and dirty, the one you chased
Has vanished, shown a clean pair of heels.
The Metro thunders through like heavy guns
To shake the waking streets. They are effaced,
And reconstructed, effaced again.
Now where are you, psychopompos?
Who'll pick up your thread or catch your train,
Who'll follow you and bear your mouldered cross
Through tunnels tight as fingers in a glove?

*

Desire again, the undersong. The lost
Children feel it in their sleep,
And turn uneasily to the wall through which
Symbols pass and cool their blood like ghosts.
My mother's family has passed through it,
Not one remains, and she is half way through.
Her brother disappears, the glove has closed
About him somewhere and dropped him in the ditch
Among the rest. The ditch becomes a pit,
The pit a symbol, the symbol a desire,
And this desire's the thread. The tunnels creep
Under the skin, the trains with their crew
Of passengers can glide through unopposed.

*

Their voices are not heard but seen, are moving
Lips and tongues. They're well behaved and quiet.
To give voice is to lip read, to construe
The contortions of a mouth, to place the living
Where the dead are, your money where their mouth is.
The body longs for touch: no words are spoken,
But sentences break up, are made new
Into fictions which will occupy the city
Like a foreign army. Their all too tangible bodies
Litter up the place, these men and women
Travelling. Her voice is underground.
Her poetry (unseen and without sound)
Lies not in pity but in clarity.

*

The Metro runs along to City Park,
That is a fact, and all along the line
The shivering persists beneath your feet.
The same with facts. It is a chance remark
That lingers in the tunnels, is embedded
In pavements, under skin or in the grain
Of your bench. She steps in, finds a seat
And is whisked off to meet my father in
A flat in Rózsa Street. His heavy lidded
Eye remembers, re-encounters. The street
Of the rose. The rest is not my business,
But a picture in a frame. Under the skin
She wears another skin, another dress.

*

22

3 Portraits

At fourteen she went skating on the river
And caught a cold. *The boys would come from far*
To sing outside my window, there were many—
Because I was attractive and vivacious.
The cold developed into rheumatic fever,
Thrombosis followed. Then would follow a litany
Of lost relations, lost names, and the brother
Who failed to love her, who was beautiful.
The town was Cluj, then known as Kolozsvár,
The district Transylvania. From this
Follows the following, expand the cool
Shadows of biography and synopsis.
Even now I know little about my mother.

*

From this and something more, the skeleton
Of something—body, city, staircase, wall,—
Which feels impressive, is part visible;
From this follow internment and arrest,
The family hatreds and the fierce ungracious
Vendettas of my childhood, and the fiction
Of history which makes up Budapest
And what one thinks of as oneself, that one
Who thinks he sees, who wears both belt and braces
In photographs, an infant contradiction,
A narrator, himself of morbid interest,
Whose scented aunt and God have settled down,
Whose eyes shut windows in the city's face,

*

A peculiar little old man of a boy,
A kind of dwarf, benevolently wise
And puzzled, deep voiced, comical almost.
He kneels under a table, his bare bottom
Sticking in the air, a bendy toy.
He swings between the doorway, opens his eyes
And thinks he sees the faint trace of a ghost
Among the coats left hanging on the hook,
Touches the piano, examines the vast stove
Which dominates the corner of the room,
Deploys his troops on battlefields of blankets,
And colonizes every possible alcove
Of his world with a vague unfocused look.

*

The early fifties: Uncle Joe's broad grin
Extends benevolently across the wall.
The boy wears a Young Drummers uniform,
(A blue tie with a toggle), shakes the hand
Of dignitaries at some parade. He is thin
And pinlike, almost cavernous. The school
Is pleased. He's learning to perform.
She works and he works. She checks his work for him
And terrifies him into excellence.
He has a line of stars in his book. They're hers.
He watches her pupils contract and expand.
Uncle Joe's moustache will shelter them.
This is the era of benevolence.

*

Her likenesses are caught on film. Her hair
Has flared into a dark corona, black beams
Of sunlight, thick, now piled high, now falling.
It hides her face. She stands on the ramparts
Of the Bastion, her teeth gleam.
Her finger can bend backwards in a curve
That is quite frightening. She deploys her arts
Of fascination as he does his troops.
She sits down on a chair
Invites him to her lap. He will deserve
Her attentions. He listens to her calling.
His father enters and pulls him through the snow.
She smokes a cigarette and parts her lips.

*

He's easily frightened; when she lashes round
In a monumental fury he keeps her sting
In the bottom drawer. It is his occupation
To bring it out at night and scare himself.
For a long time he can sit and watch her working,
And feel her warmth, and listen to the sound
Of her breathing. It gives him an odd sensation
Of belonging/not belonging, half and half.
This half and half will always seem like truth:
(I too can see him only with one eye)
He'll keep her face and others in the drawer,
With her own photographs, her frozen youth,
Her unsent letters, his unwritten reply.

*

4 Flying Backwards

The accident of being who one is.
The accident of being in a place
At one time not another. It is not grace
Of form, but grace of accident that gives
A building power, and lends the body strength,
The necessary structure to survive.
The tattered dress of fortune parodies
Our specious dignity. It makes us eat
Our words, as I eat hers, takes breadth and length
And swallows them whole. It is the street
Of Roses. It is the beautiful brother.
The things that might smell sweet by any other
Name we give, or recall by accident.

*

I have her brother's face—a studio piece
Of circa '29, and then again
Some two years later, hair cropped tight about
His delicate skull. She stands beside him, pouts
And stares, waiting for him to release
Her trapped hand. Photography, her trade,
Is this security, this collateral:
My fiction turns to sepia in its presence,
All subterfuge is instantly displayed
For what it is, a brief ingenious pattern.
And he cared so little for me if at all.
I tried to find him later but in vain.
My words for what she meant, in a general sense.

*

The uncle with the chocolate factory,
The uncle who was magistrate,
The father who travelled to the States
And worked as a labourer. The middle class
Jews of Kolozsvár are the lost history
Of which she hardly spoke. Mother's bob
Is a fashionable frame for her neat face,
Which the edges of the photograph reframe.
They bind the sepia, prevent it spilling
Across the desk, hold names
At an aesthetic distance, where, by willing,
We can work them into fictions and animate
The past, which remains forever another place.

*

But it does spill over. It is what we are
And what we see and time and again forget.
It's there in walls, in Uncle Joe's moustache
Which is the wall. The other place is here
And grows moustaches, breasts, Edwardian collars,
Wears miniskirts and co-respondent shoes.
Whatever preserves the late imperial texture
Of before, that keeps the pattern true
And cynical, expresses its regret
In the rhetoric of faded architecture,
The blatant half truth/half light of a picture,
A vulgar hybrid of fleeting, local colour
Where only light is faithfully reproduced.
The rest is reconstruction and conjecture.

*

The rest is data such as: *At the hour*
When the Germans entered Budapest we were
Sitting in the Astoria, or *The man*
Who called for us wore glasses. Or *She used*
A certain colour lipstick . . . Such power
As we retain resides in these. I build her
In meccano. Here's the skeleton.
The bare bones of the story are reduced
To ashes and a name in Golders Green,
Behind the Hippodrome, behind the station.
Recovered from thrombosis, at eighteen
She left for Budapest, an invitation
In her handbag (possibly her pocket).

*

She worked as a photographer. The war had started
But you'd hardly know it. She met my father late,
When he returned from camp in Proskyrov.
It was February, 1944.
Next month the Germans entered Budapest
And he was recalled to unit. The date,
Nineteenth of March. Facts, bare bones, the rest
Are silences. A 'safe house' in August
With father's family two floors above.
September, October, the Arrow Cross, the raid.
Her feet are clattering in the gallery,
His family are hiding or departed
And only she remains to be betrayed.

*

5 Betrayals

Betrayed? She felt and thought she was. But who
Betrayed her (if it was betrayal) and how?
Betrayal by omission was the way,
Betrayal by those she trusted. Down below,
The soldiers in the yard, the quasi-military.
They called her down. It was a minute's work.
But why was she out on the gallery
When it was far more sensible to hide?
And why did no one tell her? Who were they
Who should have done so? Why did they shirk
Their human duty? The wound was always fresh:
Even at fifty-one, the year she died,
It bored and tunnelled deep into her flesh,

*

Katona József Street. The Swedish house.
My father's family came from the North,
Moravia and Bohemia, tailors, painters,
Vendors of musical instruments, a broker,
And father's father was a shoemaker.
How much is all this information worth?
The list is endless and monotonous,
Their season's over, summers, autumns, winters:
Few made the final spring of '45.
My father had been brought up by the aunts
Who coddled me in my turn. To survive
Was an achievement, but my grandparents
Were under-achievers all of them, bar one.

*

My father's mother. Large eyed, beaky nosed.
We must have met but I've no recollection,
Except of something owl-like, something scented.
Her absence gave her little enough protection
From mother's fury. The matter was closed,
No letters exchanged. In '56 she went
To Argentina, wrote, sent messages,
But his replies were censored or forbidden.
Her case was settled. Sometimes she sent me presents,
Pale useless things, the kind my mother resented
And fanned her hatred for her. It was the hidden
Secret of my childhood, what she'd done.
Even now I don't know what the truth is.

*

My aunts (or great-aunts to be precise)
Brought father up. His home was there. The owl
Had farmed him out. When evidence
Was weighed at home this counted much against her,
And nobody replied in her defence.
Numerus clausus, nullus clausus: twice
Father was hit by laws against the Jews,
His education stopped, he worked in knit-wear,
Was twice promoted then forced out. A friend
Advised him, trained him, offered him a place
He couldn't decently refuse:
Apprentice plumber, master of the bowl.
For both of them it was an hour of grace.

*

Inevitably, labour camps. How many
Perished here: the artists, writers,
Musicians, plumbers, brothers? Escapades,
Adventures, tragedies, the company
Reduced, disbanded then recalled.
The dark eyed girl in February, back home.
Road building, retreat, escape. The waters
Close about my grandfather and fold
Over him in Auschwitz. Brief episodes
Of dire intensity, each trivial sum
A fortune lost. The dark eyed girl moves in
With father's mother, sister, baby niece.
It's lists and rosters, jigsaws piece by piece.

*

They fall together, stand and fall together.
The day the soldiers came she was alone
And heard them shouting. On another floor
The female threesome. She looks for them. The door
Is open. She calls their names, the mother,
Sister, niece. They do not answer. Where have they gone?
Where are they hiding? Nothing. Not a sound.
She wanders out, is spotted. It is fear:
Fear of discovery, fear of strangers. It's done.
They have not answered. Someone shouts, Come down!
And who is it pretended not to hear?
The rough voice rises. I speak for another,
And buy my ticket for the underground.

*

6   In Her Voice

Like a girl listening behind
A membrane for a football or the knock
On a door, the rattling of the blinds
In a secret room of her head, I unlock
My eyelids pressed
Against a darkened window in a house
Whose eyes are asleep,
And quietly get undressed.
I have a thousand eyes to guard my neatness
Against the gods of lust, who quaintly creep
To music as they blind me and I wake
To hear the scuttling of a mouse
Here on the fourth floor at one o'clock.

*

I was on the fourth floor when the yard
Filled with uniforms and we were called
To order, and I ran into the flat we shared,
The old woman, her daughter and the child,
And all was empty. I whispered their names
But they did not answer to their lasting shame.
They should have answered me out of the pit,
Like any prompter from his own hell-hole,
But they closed their mouths to my pitiful dole
So I went down and here's the end of it.
Those men have strolled at ease about our yard
But God will grant them their reward
And punish them according to their lot.

*

They took me in their wagon, up the street
I used to walk, with all its empty faces
Staring from the coigns and pediments.
Great figures started from the roof in tents
Of stone and tiling, forgetting the discreet
Darkness of their long discolouration,
And tiny figures flitted by the bases
Of the portico of the academy.
My mind showed little sign of occupation:
All life was going on outside, upheld
By the conventions of the weather,
My tenants were expelled
To railway stations where we lost each other.

*

They put me on a train, east, west or south
And we rode off in our different directions,
Myself, my body and my heart. My eyes
Were saying something to my open mouth,
Which had remained open in surprise
And every passenger had his own questions:
My nose asked, what's the smell?
My fingers wondered at the touch of cold,
My hair was busy interrogating the wind.
We were all agog to know the world at last
As it knew itself but never before had told
Anyone. Nor did I mind
Whether this was heaven, earth or hell,

*

As long as we were moving in the air,
As long as the city barked its orders out
Through doorways I imagined everywhere
And heard the porters shout
Behind closed eyes and behind the narrow wall
Of my most valued multi-storeyed skull.
　　But they told me no great truth or if they did
I have forgotten it. It was long ago
And I have doubts whether such a truth
Exists at all, as something we might know
Or understand, I have my hatred
Which is proof that something happened in my youth,
And the house itself has not yet been blown down,

*

My body is still standing. The wind blows through it
Like a language, of which not a word
Is what it seems, and yet it survives.
The train is rushing past the fields and woods
Of all that was. The words renew it,
Rephrase its truths and falsehoods.
Behind the thinnest of walls a city thrives,
The empty buildings, the unfurnished,
Whose history remains unfinished.
　　I rush out to the gallery, alone
And watch the soldiers massing underneath,
My brothers all, their justice bone for bone,
Their eyes are my eyes, their teeth are my teeth.

*

7 *What should they do there but desire*

Disorientation, loss: the doors that close
Just when you think that you have gained your entrance.
A glimpse of hallway, hat-rack, mirror, more doors.
Beyond the doors and on the left perhaps
A window giving on to a neat yard
With trees and flowers. Straight ahead of you
A lift-cage dressed in iron broderie,
A smell of coffee brewing, an envelope
Slit like a wound, the darker recesses
Of sitting rooms, momentarily opened.
What troubles me is the uncertainty:
Is this really a valuable darkness,
Or am I part of the darkness that's locked out?

*

The wind is scrabbling at the glass—perhaps
The trees are wanting to be let in.
The branches say nothing
Expressing only an incoherent thirst
For music, a music so violent and awful
That it can only leave them waving their arms.
Imagine the cellos sprouting dark green tongues
And moaning softly of their lot; their past
Of growing, cutting, hewing, shaping
To this one point of supreme helplessness.
What's eating them? And yet it's good to be eaten,
To become the food of passion and to feel
The stomach rise in suicidal independence.

*

The wind stands in high places and looks down
And comes out at your arsehole and your mouth.
I do not speak now as a lady should,
Not even as a woman, but of parts,
The one dissociated from the other.
When I think of you I only see your head,
Sometimes a hand. The wind runs through your fingers
And cools the blood to a blue stream of air,
And when I hear it scrabbling at the glass
I'm filled with pride and understand the light
That leaps inside me and across the table
To reach out for some part, a head or hand
Or thigh or foot or armpit, something of you . . .

*

Some years ago I met a man by chance
In a foreign street. I had not seen him since
The time when I last saw my brother
Of whom I carry about this sepia
(I seem to be my brother's only keeper)
Which is paler than he was. I blame the weather
For his fading and our having grown estranged.
He was handsomer than any man I knew,
As handsome as a woman could desire.
    There was a policeman once who doffed his cap
And showed how in the lining he arranged
His family in tiers of small brown snaps.

Photography, I need you. Freeze me too.

*

Even here there are shadows of places: serene,
Impassive, idiotic, undemanding,
Without bitterness or rancour.
We are travelling in darkness, standing
On each other's feet but at one remove.
The door of the wagon rattles a pale music.
An elderly man is sick in the straw.
A child clings to my thigh. Two grown men kick
Each other in a fury, or try to gnaw
A third man's head. What do these things prove?

It is the peculiar happiness of buildings
To be witnesses. Here are the stones and mouldings,
The molten forms of clinker.

*

Dear brother, I have talked to everyone
But no one knows you. I am sitting in
A wooden hut, rather like a kennel.
You're well away from here. A woman kicks us
As she passes. I do not trust the women.
The men we're used to, they are what they are,
The usual sheep, but I'm a woman and
I know that pitch of the heart,
Am living in it. Who are they paying back?
Their elder brothers? Wherever you are
This non-existent paper will locate you,
In the angle of the wood, the nursery,
Or up the cherry tree with its sticky black cherries.

*

Here's Ravensbruck. I stop dead at the gate,
Aware I cannot reach you through the wire,
I cannot send you poems or messages,
No wreath of words arranged across blank pages,
No art that thrives on distance and desire,
But can't cope with fulfilment, that writes white
When happiness breaks out, that lights a taper
On a frozen lawn and bounces off the stones
Of hard luck. The dead have no use for art.
You might as well bring on the tongs and bones
As chamber music (Schubert's great quintet).
Not all the white ink in the world can set
Their coming through, their verses on black paper.

*

And if I bring you here and push you in
It's only because I know you once came out.
You cross the black bridge thus. *Ich bin allein,*
Ich stell die Aschenblume ins Glass voll
Reifer Schwärze, deep into your mouth.
And if I attribute to you desire
It is to replace what was voluptuous
In bodies full of warmth, *das aschenes Haar*
Which is also mine. I wait outside your school
Of hard correction, mouthing words too soft
To bear a lasting mark, a feminine tongue
In my head. I float on my own craft,
And try to write the half dead a live song.

*

Dead grandfathers, dead grandmother, dead uncles.
Item: to my children, All the aunts
Their grandparents can muster, this bequest
To be taken by them for granted. Tender plants
Turn vast familial trees in paradise,
Which is nothing else but superfluity,
Where every woman has an extra breast,
And every generation's spoiled for choice.
A balloon floated past our window and almost
Touched the bricks. Small green leaves covered
The trees. So wrote one Nelly Toll in Lwow.
Superfluous in base things, we are lost
In distant towns whose names sound much like Love.

*

See, in this drawing a girl is making Lwow,
Her mother and she are playing dominoes.
The sunflowers are growing in the shadows.
Small green leaves cover the trees. Above,
A balloon floats past the window. *I visited*
The children on paper. Paper of deep black
Is lightened by her painting. Even the dead
March cheerly in their prison, and look back
On paradise, and know that God is Lwow.
Ich stell die Aschenblume ins Glass voll
Reifer Schwärze. The camp choir sings a mass,
The camp dogs chew their bones. And in the glass
A brilliant ash-grey flower for Nelly Toll.

*

She tolls me back to the bleak scene before
The entrance. The women march to the factory.
Their wooden shoes are tottering on the ice;
It sounds like someone knocking at the door.
Thirty years on the knocking hasn't stopped,
But now your heart wears out its battery,
Is running down, its tick-tock, less precise,
Is more like memory, which soon is lost
And drifts above the garden in fine dust.
Like Lili Marlene I wait outside the gate,
A lamplit watchdog expecting no returns.
Imagine my surprise when you walk out.
The crematorium waits, the oven burns.

*

This is our lucky day, like every day.
The white ink settles on the page like snow.
The sunflowers are growing in the shadow.
The crows are circling looking for dead meat.
Your hair turns into flowers streaked with grey,
I put them in a glass. The room grows warm
With soft grey flowers, responds with its own heat.
The pillows, blankets, curtains are in bloom
And open into grey. The grey flies swarm
About the lamp. This is our mortal room.
A train is arriving at an empty station.
A voice is speaking, but it isn't mine.
The passengers are spilled across the line.

*

9 Fraternal Greeting:

Beauty and terror, just enough to bear:
The Rilkean brother, a little lower than
The lowest angel whose indifference
Is murderous to those who marvel at him
And expect him to return love like a man.
Disdain is an improvement on despair,
And hatred perhaps a kind of confidence
Which can be shared like intimacy,
Unexplained antipathy or dim
Persistent loathing. No analysis
Avoids abstraction. For some there is no physic
Or improvement. When instructed to kiss
His baby sister my uncle was violently sick.

*

And so there's guilt, guilt and indifference:
He in his turn became idealized
As all disdainful gods are. That is why
They are gods. And this was no pretence:
He could be human, true, but not to her.
She called him 'freedom fighter', 'partisan'
Once she had lost him. For a god to die
Is only to gain in potency, to rise
A few clouds higher—and it might have been all true,
Although I heard he was a prisoner
And laboured, like my father, like a man,
And then might have been shot, aged twenty-two,
Somewhere in Slovakia perhaps.

*

Perhaps. And yet he simply disappeared.
Tall forehead, dark hair, full and sensuous mouth,
Intense, intelligent. But to be sick
At touching her? Are gods sick at our touch?
It's possible of course. The bright one's beard
Is sensitive. It bristles at our youth
And emptiness. To him, we stink of pitch,
Are lymph and chyle and hatred. When we raise
Our holocausts to him he looks away.
He's not pleased by the smoke of sacrifice,
Or tawdry festivals or holy days,
Is tired of them. The gods have seen so much
Of fire they begin to turn to ice.

*

It's not that they mind the flattery: the smell
Is what offends them and they cannot help it.
So he's a god too, and whatever pit
His murderers interred him in he rose
And faded, entered other realms like hell
Or heaven. And she in her hell yearned
For his beauty and affection all the more.
Desire and pain. Around her bodies burned
In their own fevers or behind the door
That was always round the corner. I propose
A yard, a hut, a fence, a row of beds
And shins and shanks and ribs and collarbones,
And one familiar among shaven heads.

*

Those burning babes, visions of Christmas day,
The small photographers, provincial towns,
Perversities, distortions. They move down
The escalator, spread along the platform.
The train arrives and takes them. More keep coming.
Each face desires another. They pray
With a look, communicate by grimacing.
Each one of them is in some uniform
Of obsolescent dullness. Each bears a name
On the collar or the sleeve. Their names are numbers.
Above their mouths a single flickering flame
Sustains their spirit, and like spirit, it burns
And dances and reduces them to cinders.

*

Dear brother, I have talked . . . the voice is distant
As the past it conjures, as the little boy
Under the table . . . *but no one knows you.* To know
Is not to see or understand. The grey
Fly hovers at the curtain without knowing
That any particular thing is so
Or otherwise, but his drone remains insistent.
The walls keep mouthing at us with their doorways,
The trains keep coming and going.
The ghosts must pass through the walls alone,
Take on the character of stone,
Seek out the angle of the wood, the nursery,
Climb up the cherry tree with its sticky black cherries . . .

*

10   *The little time machine*

Burnt offerings: a little bonfire shivers
At the far end of the street, all rags and card
And insignificance. A wheelbarrow
Is propped like an old man kissing the pavement,
A stiff frock coat, the mud on the wheel his beard.
The flames leap and fall in rapid rivers
Of light, a confusion of elements.
I see small fires along the narrow
Passages between main thoroughfares.
The heart, the eyes and passions maintain
Their vigilance. The holocaust goes up
In smoke. Somewhere a soldier prepares
To set fire to fine details on a street map.

*

The map is always burning. Its consumption
Is conspicuous enough, imagined cities
Of fugitive colour, changing light on tiles,
Faces at windows, hands at doorways, feet
On trams and buses, clothes in smelly piles
In empty hallways, the sonorities
Of gossip and greeting. My friends and I meet
At restaurants, complaining of hard times
In the benevolence of an August night
That smiles on our children. We are an exception
To the rules of sleep. Our children will sleep light.
After the fireworks we tell old jokes
And pay our debt to history with rhymes.

*

The city dreams an island. It has always
Been here, stacked on its mound of days
Lapped by cold sea, pickled and saline,
Wearing, breaking off. Hard water furs
The kettles, houses fall, rejig the shoreline,
Everything is continually in friction
With the wind off the sea. The women with scarves,
The men pottering in sheds, seek protection
In distance, the insularity of it all.
Sad, great, shaggy country. The soldier hears,
Takes aim and fires but misses. Foreign flotsam
Adheres to the feet of piers by decaying wharves.
The ferries shuttle. Waves crack on the wall.

*

The crack of a gate. Time opens backward to
A heap of pebbles suspiciously like bodies.
The wind whistles through trains whose nightmare crew
Of passengers have fallen quiet, stopped
Their grimacing and squealing and have dropped
Where they stood, dropped off to sleep at last
In broken postures, parodies
Of grace, recumbency and carelessness.
It is only by imagining the trains
That I can enter the gate, walk across the field,
And wait for the signals to announce the express
Europa. Its carriages are sealed,
The wheels go rattling over broken chains.

*

Too long rejected, we meet up in the street
Below a lamp post, yellowed as old papers.
What news? we ask each other. Our faces
Are the cut-out shapes of childhood, full of creases
And torn edges, smudged and circled
In soft chalks. We've brought along with us
Giraffes and elephants in a discreet
Procession, with dolls and packs of cards, and pieces
Of furniture arranged in packing cases,
Nothing but dust and detritus.
This is the news, hot off the world's press.
It's late at night, you say. We are light sleepers,
I reply, our sleep is a kind of emptiness.

*

Somebody has escaped at last. Somebody gets married,
Has a child, another. Somebody remembers
Someone else or something, certain numbers,
Certain streets and faces. One is worried
By forgetfulness, another by clarity.
Someone is not sure they should be here.
   Down into the Metro, down the stair:
A drunken woman's weeping on a bench,
Another's sitting in a pool of water,
The horrible familiar stench
Of loss. A fat policeman nudges
At them. The crowd skirts round the edges
Of the frame, spreads out into the city.

EN ROUTE

1 MY NAME

A voice in another room calls out a name
you do not recognize but know is yours.
So many ways with so few syllables
    that you must claim
along with clothes as dead man's metaphors.
Your very heart comes to you bearing labels:

a citizen, a Jew, bourgeois, brunette
or invalid, and each speaks differently.
your parents with their formal admonitions,
    the aunts who pet
and tease you, the lovers who will gently
bend your ear with their murmurs and petitions;

a coin, a patina, a doctor's file,
accretions of diseases, school reports,
employer's references, catalogues
    Shakespearian style
where murder lands you in the kangaroo courts
of popular feeling. You are one with dogs

and cats and household familiars like germs.
Your shaven head, your wooden shoes, your bed
with its number; your fleas, your lice, your store
    of serious worms
that wait on both the living and the dead
whose numbers you will swell. I keep the score

and tell the days religiously on which
my name took on new meaning with a hat
or spoken word. I keep up with the fashions
    and have grown rich
on music composed of lists. I have sat
in contemplation of my ruling passions.

I tend the days like tiny photographs,
each clearly labelled with a name not mine
but valuable: features, smells, a profile,
    a mouth that laughs,
a hand in my dead hair, a lip to sign
and seal me, and your vacant missing smile.

The missing items haunt me. Ghosts that fade,
too lightly pencilled in, underexposed,
time is always threatening to efface them.
    My hands have strayed
in filthy water. The river's doors have closed
on teeth and bones, not mine. I cannot place them.

2 THE LOVE OF WINDOWS

1

I love the height of windows—they are bodies
at attention in their black and blue,
are blinding truths or lies, or silent studies
in deception, something seen through
once and then again, they frame us, me and you.

Your eyes are windows too, a blank display.
I cannot make you out: the life you lead
turns on no sudden light to make night day,
is only words too small and faint to read
like something in a contract or a deed.

2

Light strips down at night to ghosts of itself,
flung across the room like an old jacket,
so thin it has no weight at all, grey as a wolf
and small enough to fit inside your pocket.
Since boys have pockets full of useless stuff
I have been certain there is room enough

for light and ghosts. For both. And there are boys
I see across the pavement whom the wind
seems to carry, and a boy who plays
with his disgust, who's neither mine nor kind.
A girl will find such hardness, swollenness
offensive as an unbecoming dress,

Yet feel a numbing tenderness towards
these martinets with secret business toward.
She'll try to hurt them into words with words
but find at the kill her own self being gored
by (let us call it) love, romance (or such).
Her words are not enough. Or far too much.

3 GUARDS

I remember the windows in the evening;
five grey guards above the bed, their thin
pale uniforms were not assuring,
what monsters might they not let in?
Might they not all float in unobserved
and let loose the damnation we deserved?

Wry faced clouds and empty dinner plates,
cloths hung on the pipes under the sink,
webs in the attic, shattered slates,
the faded ink
on newspapers whose eyes dissolved in dots
and yellow patches,
a scumble of coal dust and dead matches,
the earth round flowerpots.

I screw up my eyes in the dark, and see
your open brown eyes staring back at me
with all the artlessness
and heartlessness
of childhood, until the tears come.
The guards as usual are waiting, playing dumb.

4 PIGEON CHESTS AND ALARM CLOCKS

I see the apple blossom puffing out its pigeon chest,
I see the clouds, dressed like doves, in down,
strips of corn and vines, a field of sunflowers
    goggling at the sky.
There are farms beyond the edge of town,
churches buttoning their jerkins neck high,
    and pines like thick brown showers.
I feel a heaviness, a dull weight on my breast.

We're off, we're off! Do you remember the excitement
when our parents visited their friends
in the next town? They bought us chocolates
    and bags of sweets to clutch
and took us to the station. I can hear the sounds
of tickets being punched, kids crying. I watch
    the hedges, banks and gates
all whisked away, a landscape of retracted statements.

The big cities are waiting on the line. Their backs
are turned to us. They plan their moves alone
without our help. Their streets are like cracked glass
    or folds in old dry skin.
In either case they crack. Their walls are bone
and just as brittle. The very air is thin.
    When we wake up we'll find the scenes we pass
wound up and ticking at the windows like alarm clocks.

5 FATHER IN AMERICA

When father went off to America
he wore a white suit. Later he shaved his head.
Our ancient town burned white throughout the summer
and white flowers blossomed in the cemetery—
the flowers of course remain though he is dead.

It was everybody's most idyllic picture—
white suits against pale brick or amber corn,
and gentlemen and ladies in such posture
with parasols and petticoats in pastoral
benevolence before his head was shorn.

I fear shorn heads—I touch my own skull now
and feel skin pimpling in between the roots,
with father's skull beneath. I feel it grow
progressively more bulbous under mine,
his brain is still developing new shoots.

I feel, but know that feeling isn't knowledge.
The glass distorts in the amusement park,
your skirt blows high, you cross a swaying bridge,
you hear a scream, you stand before the mirror
and face your masters in the gathering dark.

I wish this train were going elsewhere but
a wish is powerless. The skulls appear,
one on top of another. The doors are shut,
and I'd be lying if the truth were told
on picture postcards, wishing you were here.

6 A SOLDIER

after Károly Escher

A young man with two flowers in his cap
Has turned away across the platform
To move towards two women wearing headscarves.
He is the country I am leaving.

He is beautiful, a beast decked and garlanded,
He stands gently and placidly, tall, slim,
Melancholy, prepared for sacrifice,
A peasant soldier, simple as they come.

Death has half closed his eyes
Ready to devour him at a blinking,
Behind his head the blur of a wagon pulling out.
He seizes one of the women, embraces her,

Presses himself against her.
As we depart I am tempted to shout
To attract his attention. I can only guess
The occasion of his death, his tenderness.

7 BORDER CROSSING

You leave one body, enter another, thinner than
The one you wore. Having nothing to declare
The customs do not bother you. You pass
To other gravities, no longer man or woman,
But neuter as the clothes you wear
As thin and transparent as glass.

In the glass you see anatomies,
Bacteria and germs in broken places.
You see the future in slivers and shards
Faint, farcical lobotomies.
I try to discover my disease in traces
Of tea leaves, life-lines, livers, tarot-cards.

Impossible to read the auguries:

The future waits on fiercer surgeries.

I think of the colour of corn, each stem a stork
on one pale pinkish leg. The fields are full
of one-legged birds, brown trunks and burning heads,
nature's paraplegics, stunted, cropped,
the Bosch-men and the peasants out of Breughel,
the playful demons round old misers' beds.
I am not superstitious but go to work
on heaven to find the universe has stopped.

The bogey men will get you in the end,
the story goes, or else, *then there was one* . . .
The deep blue air has many tales to tell,
O children, yet you cannot help but laugh.
I hold your hand beside the table. Sun
flashes from the bright machine. We swell
into a prominence. Till then I send
my love, the light within a photograph.

9 POUDRETTE

Inodorous and inoffensive poudrette:
our dust and ashes and our waxy bones
which slip so easily into a basin to wash the dead
cleaner than clean, to cleanliness as never before.
You feel the round shapes in a woman's head
and realize the human race is clones
and talk, a mask of fragrant etiquette
to cover up the smell of decomposition.

I like the story of Caravaggio's whore
playing the Virgin Mary; I also like
the Raising of Lazarus by Giotto,
all that foulness turned to good use, like a motto
based on others' suffering. We have a mission
to convert the heathen and to slake
their thirst for knowledge. Look, we may say,
this is what we have to offer you.
We are the Fathers. Here is the Gospel of the New
which we have brought down from Heaven this very day
to proclaim unto both the gentiles and the Jews.
Henceforth ignorance will be no excuse.

I imagine the obsessive tidiness
of those who love and fear for each other:
girls who straighten ties, men who address
themselves solicitously to their lover's wardrobe,
all uxorious husbands and doting mothers;
and I imagine the obsessive tidiness
  of two people curled up in a bed
    asleep or perhaps dead,
a married couple, sister and brother.

Their breathing is their wardrobe; they
are clothed in air, the fabric of their suit
has turned to dust, evaporated in a play
of shadows, collapsed like a net.
Under the flesh their bones assume command;
the tendons, ligatures are shrivelled fruit,
  old apple cores and twisted pears,
    a spindle of fine hairs;
their scalps have become dry sand.

Perhaps then, when they are nothing but bone
the urge for neatness grips them. They hold
each other together, arrange their unknown
components in secret patterns; link, omit,
dispose. The dry tuneless piano in the ribs
has lost its hammers, but some tune unfolds
  within the vertebrae,
    a kind of harmony
is understood by hardened ears and lips.

A GREEK MUSÉE

When I look at my room I see powder. Life as a footnote
to unwritten literature. The chair with its thick varnish
picked up at a junk-shop, heading for a junk-shop,
is preparing, even now, to vanish.

A few thousand books gathering dust and amber
and half the books not read.
Literature is this torn old pair of slippers.
The plaster flakes and weals above my head

continually aspiring to the condition of literature,
the facets of a crystal. I listen to a record
knowing every voice on it is dead but breathes
volumes into my chaotic word-hoard.

I inhabit a communal Musée
des Beaux Arts where all things learn through error,
perfecting their falls from grace. I read the papers
for anthologies of terror.

And look, a shepherd watches a child fall
on to Greek soil, followed by the mother, followed
by a man's leg, followed, it seems, by the sky.
Literature is Chicken Licken's fellow.

When I look at my room I see Greece. The bloody Gods
are resting on my two seater settee,
modelled on Habitat and falling to pieces now.
All will be patched up in God's new city,

all will be literature, as perfect as the armour
in the basement of the Fitzwilliam; the plates,
the pots, the pictures and samplers, and the drafts
of Auden, Spender, Tennyson and Yeats.

THE OLD NEWSPAPERS

I discovered Atlantis when I drowned.
It was a pleasant going down. My gutturals
Were smoothed out as the sea rose higher
With all the colours of petrol.

Old warehouses of stuff from '51,
Left over from the Festival of Britain,
Faces from Ealing comedies, dead Sellers,
Dead Hancock, a benign pattern

Of cheerful decay. I heard the manic laugh
Of Secombe and saw the exaggerated curve
Of a girl's rump slipping behind the piano
Crying, *I am what you deserve!*

(She squeezed her breasts to emphasize the point.)
Boxed gardens bristled with box and swam in a slough
Of their own despond. Ducks, geese and chickens
Would regularly plough

The soil to lead and silver. Up they drifted
And rose too, the dried kit of winter games,
Flakes of dead grass, the smell of newspapers
With urgent crepuscular names:

The Evening This and That, the terraces
Where headlines lost their vim, sodden
And mushrooming with fringed clouds in a cellar,
In a rhetoric forgotten

By Cassandras, Crossbenchers and Spot the Ball.
My face dissolves in print, a ticket
For the underground gone under.
I feel the weight of the pound in my pocket,

And know it to be the same, a green sea lettuce
Wrinkling like a face, its lips pursed for the kiss
Of fossilization, knowing the softest leaf
Must shrink and solidify to something like this.

PRELUDES

1

All evening I kept running over leaves—
their small dry hands were spiders scurrying
until I broke them. They were in a panic,
some cataclysm might have overtaken
the whole leaf population as I started
down a B-road behind the motorway.
East Anglia, Home Counties, darkened twisters
of hills, punk hedges, sleek old villages
in lamplight, shops with lawnmowers and chainsaws,
the distant sky a television glow
between high trees. I thought: this is my prelude,
the beginnings of a gothick fantasy,
half-pulp, half-mysticism, school-of-Palmer,
the understatement of an English landscape
whose skin is tight and heavy, lumps of shine
receiving a faint covering of moonlight.

2

It wasn't like this in the soft green books
of Dickson Carr, Ngaio Marsh and Christie.
The libraries, the dens, the inglenooks
Have only ghosts to fear,
Though when you look the moon is far from clear
And the moor is slightly misty.

How can you take a murder seriously
When everyone's in costume? Bumbling inspectors
Stage their re-enactments with mysteriously
Poor results. Old cars
Are found abandoned under field of stars
Rusting away with harvesters and tractors.

A lack of motive, boredom, pressure, panic.
A television shivers into light
Against an empty wall, with its babbling manic
Insistence on being heard.
Whatever was going to happen has occurred
And is repeated hourly through the night.

A single man could creep out of the hills
and, moving purposefully, disappear.
A smell of petrol and an unlit car
on which the well-intentioned moonlight falls
with its mandate for restoring law and order.
The latest pastoral involves a murder.

Discarded anoraks, kaguls. The killer leaves
a trail of functional overgarments, camps
and boots, four empty tins from petrol pumps,
and, underneath the hedge, a pair of gloves,
reminders of the hands that broke from wrists,
a flaccid emptiness formed into fists.

Her hands, or someone's. As the leaves shake free
of barrel-loads of rain he sees her hands,
sole presences in stadiums, a stand
of favours, green and purple like the sea.
A pub winks in the distance. Music sweeps
across the fields: the damp grass bucks and dips.

He goes to ground and finds a hidden nickname—
Some rural creature, rat or mole or fox
which show him only their inhuman backs.
The earthworm fits him and invites him home
to supper where they share a piece of turf
and someone strangles him with his own scarf.

4

A lorry bent its head and seemed to graze,
A tall brick house backed on to CRODA GLUES
And the grass was rotting baize.
The fields of rape hummed at a ragged sky,
Long strips of surgical tape were laid across
Young lettuce. I watched the hulks of sleeping sheep emboss
A pasture, having dragged along the hedge
Their trail of dirty candy floss.

A Pakistani read *The Yorkshire Mart*,
A headless windmill guarded British polder,
A shaft of sunlight opened like a guillotine,
And a white horse drank his own reflection
In a muddy field. Everywhere the green
Retreated into dampness, growing colder
And colder, to a sub-zero diction
Of bulletins that stirred my foreign heart.

BURNS NIGHT BY THE DANUBE

In memoriam 1956

1

A time for slogans: know them well.
Would any of them ring a bell?
You'd recognize them by the smell,
Proust's madeleines.
They tease your tongue and cast their spell,
Invade your teens.

Unreal applause, the names on tick,
A twitching body politic,
The headaches and the waiting, sick
With apprehension:
Before you rise you feel a crick.
Pre-menstrual tension?

You read your article and smile.
Should you have frowned? It's all on file,
Your Brief Lives brevities beguile
Astute inspectors
Of bric-à-brac, who scan for style
With lie detectors.

If they prick, do you not bleed?
Fertility is more than seed,
A broken egg is what you need
For omelette:
Your own two may well interbreed,
Intact as yet.

Then rhythmic clapping. It's the clap.
A name in a gazette or map
Is quite sufficient to unwrap
Your furled-up flag.
Does any bull perceive the trap
In a red rag?

Who can't afford to lose their grip
Develop a stiff upper lip.
A fellow student quotes a clip,
Something or other
Concerning you. Who let it slip
About your mother?

It's Mother Nature, Mother Church
And Mother Naked. Bring the birch.
The shameless hussies who besmirch
The holy altar,
Must first be rooted out. You search
And dare not falter.

Don't count the hours. The hours don't count.
Don't count your wages. The amount
Is relative. What's paramount
Is eye and ear:
The mouth or pen the only fount
And source of fear.

2

October: funeral and feast.
The corn is in, the plot is leased,
Let earth rejoice with man and beast,
In conversation:
The food is good and the deceased
Is a sensation.

A few days in ten thousand find
Our Nevsky Prospects packed and lined
By dazed and vaguely happy, blind
Tiresian toilers:
Tomorrow they'll be cribbed, confined,
Back at the boilers.

What's never high cannot be brought
Down with a bump, is simply caught
Short in public places. Fraught,
And prone to panic,
Its one response is highly wrought,
Morose and manic,

A comedy, and therefore tense
As taxes or as common sense:
Relief, as always, is immense
At every prat-fall.
The villain has his impotence,
His feline cat-call.

But everyone gets walk-on parts
In history. The action starts.
The scenes are shot. Bring on the carts
Remove the dying.
Ill fortune flings her fatal darts,
The fur is flying.

If one could only keep all these
Safely between parentheses,
Initials carved in civic trees
To spite mere nature:
Proclaiming in telegraphese
To the vast future:

Specifics and particulars
Are everything. The burned-out cars
Reveal their names. The city's scars
Are told and entered
In the records. Death appears
To have repented.

The children lying in the street
You will remember, and the feet
Protruding from beneath the sheet,
A shoe or dress.
Your own tall room, the dining suite.
The drunkenness.

OXFORD POETS

Fleur Adcock

Yehuda Amichai

James Berry

Edward Kamau Brathwaite

Joseph Brodsky

D. J. Enright

Roy Fisher

David Gascoyne

David Harsent

Anthony Hecht

Zbigniew Herbert

Thomas Kinsella

Brad Leithauser

Herbert Lomas

Derek Mahon

Medbh McGuckian

James Merrill

John Montague

Peter Porter

Craig Raine

Tom Rawling

Christopher Reid

Stephen Romer

Carole Satyamurti

Peter Scupham

Penelope Shuttle

Louis Simpson

Anne Stevenson

George Szirtes

Anthony Thwaite

Charles Tomlinson

Andrei Voznesensky

Chris Wallace-Crabbe

Hugo Williams

also

Basil Bunting

Keith Douglas

Edward Thomas